HOW TO CONTROL THE TUBERCULOSIS AND HIV/AIDS DUAL EPIDEMIC.

Philip Anochie

HOW TO CONTROL THE TUBERCULOSIS AND HIV/AIDS DUAL EPIDEMIC.

Published by

Philip Ifesinachi Anochie

Research Scientist

TB/HIV/AIDS Research Group.

Philip Nelson Institute of Medical Research.

INTERNATIONAL EDITION.

HOW TO CONTROL THE TUBERCULOSIS AND HIV/AIDS DUAL EPIDEMIC.

Published by

Philip Ifesinachi Anochie

Research Scientist

TB/HIV/AIDS Research Group.

Philip Nelson Institute of Medical Research.

Telephone: +2348140624643, +2348173175179, +2348166582414.

E-mail: philipanochie@gmail.com, OR philipanochie@yahoo.co.uk.

Contents

ACKNOWLEDGEMENT

The Authors gratefully acknowledge the help of GFATM, WHO, UNAIDS and STOPTB for their support in writing this book.

SUMMARY

This book contains the novel approaches and innovations on how to control the Tuberculosis and HIV/AIDS dual epidemic. AIDS kills more than 8000 people every day worldwide. More than 5000 people die from TB every day.

HIV-positive people can easily be screened for TB; if they are infected , they can be given prophylactic treatment to prevent development of the disease or curative drugs if they already have the disease. TB patients can be offered an HIV test; indeed , research shows that TB patients are more likely to accept HIV testing than the general population.

This means that TB programmes can make a major contribution to identifying eligible candidates for ARV treatment.

Many people , including those living with HIV, do not know that TB is curable. This is a basic message and one thing that is failing to get through. We should know how to get the message to policy- makers and funders that more drugs are needed for both diseases. It is not "either/or', there is no point providing ARVs only for patients to die of TB for lack of TB resources. Similarly , TB efforts alone are insufficient.

HIV/AIDS is dramatically fuelling the TB epidemic in sub-Saharan Africa, where up to 70% of TB patients are co-infected with HIV in some countries. For many years efforts to tackle TB and HIV have been largely separate, despite the overlapping epidemiology. Improved collaboration between TB and HIV/AIDS programmes will lead to more effective control of TB among HIV- infected people and to significant public health gains.

CHAPTER 1

INTRODUCTION

AIDS kills more than 8,000 people every day worldwide. More than 5000 people die from TB every day. TB is the leading killer of people infected with HIV. TB causes at least 11% of AIDS deaths and possibly as many as 50%. Up to 50% of people with HIV or AIDS develop TB. Worldwide, 14 million people are co-infected with TB and HIV- 70% of them are concentrated in Africa.

In some regions of Africa, 75% of TB patients are HIV-infected. TB can be successfully treated even if someone is HIV-infected. In a given year, people living with HIV are up to 50 times more likely to develop TB than those who are not HIV-infected. Treatment of TB can prolong and improve the quality of life for HIV –positive people but cannot alone prevent people from dying of AIDS.

If TB is left unchecked in the next 20 years , almost one billion people will become newly infected , 200 million will develop the disease , and 35 million will die of it. More people are dying of TB today than ever before. TB is the biggest curable infectious killer of young people and adults in the world today. TB is an opportunistic disease that preys on weakened immune systems.

In most eastern and southern Africa, where combined TB and HIV prevalence rates are the highest in the world, only about one in three TB patients currently receive a full course of TB drugs.

In Africa as a whole , the number of TB cases is rising dramatically, by 4% a year. Half of all new global cases of TB (4.5 million of 9 million) each year are in six Asian countries- Bangladesh, China, India, Indonesia, Pakistan and the Philipines.

It is estimated that one-third of the 40 million people living with HIV/AIDS worldwide are co-infected with TB. People with HIV are up to 50 times more likely to develop TB in a given year than HIV-negative people.

Another aspect of the resurgence of TB is the development of drug-resistant strains. These strains can be created by inconsistent and inadequate treatment practices that encourage bacteria to become tougher. The multidrug – resistant strains are much more difficult and costly to treat and multidrug – resistant TB (MDR-TB) and Extensive Drug Resistant Tuberculosis (XDR-TB) are often fatal. Mortality rates of MDR-TB and XDR-TB are comparable with those for TB in the days before the development of antibiotics.

With effective treatment, TB can be cured, HIV managed, and lives saved.

CHAPTER 2

AETHIOLOGY AND TRANSMISSION OF TUBERCULOSIS

Tuberculosis (TB) is a disease that usually attacks the lungs but can affect almost any part of the body. A person infected with TB does not necessarily feel ill – and such cases are known as silent or "latent" infections. When the lung disease becomes "active", the symptoms include cough that last for more than two or three weeks, weight loss, loss of appetite, fever, night sweats and coughing up blood.

TB is caused by the bacterium *Mycobacterium tuberculosis*. The bacterium cause disease in any part of the body, but it normally enters the body through the lungs and resides there.

TB is spread from an infectious person through the air. Like the common cold, TB is spread through aerosolized droplets after infected people cough, sneeze or even speak.

People nearby, if exposed long enough, may breathe in bacteria in the droplets and become infected. People with TB of the lungs are most likely to spread bacteria to those with whom they spend time every day- including family members, friends and colleagues.

When a person breathes in TB bacteria, the bacteria settles in the lungs. If that person's immune system is compromised, or becomes compromised, the bacteria begin to multiply. From the lungs, they can move through the blood to other parts of the body, such as kidney, spine and brain. TB in these other parts of the body are called extrapulmonary TB and is usually not infectious.

CHAPTER 3

TREATMENT AND CONTROL OF TUBERCULOSIS

TB can be cured , even in people living with HIV. DOTS (Directly Observed Treatment Short Course) is the internationally recommended strategy for TB control.

DOTS treatment uses a variety powerful antibiotics in different ways over a long period to attack bacteria and ensure their eradication. Treatment with anti-TB drugs has been shown to prolong the life of people living with HIV by at least two years. It is important that people who have the disease are identified at the earliest possible stage, so that they can receive treatment , contacts can be traced for investigation of TB, and measures can be taken to minimize the risk to others.

However, some strains of bacteria have now acquired resistance to one or more of the antibiotics commonly used to treat them; these are known as drug resistant strains.

It is estimated that one-third of the 40 million people living with HIV/AIDS worldwide are co-infected with TB. People with HIV are up to 50 times more likely to develop TB in a given year than HIV-negative people.

Another aspect of the resurgence of TB is the development of drug-resistant strains. These strains can be created by inconsistent and inadequate treatment practices that encourage bacteria to become tougher. The multidrug – resistant strains are much more difficult and costly to treat and multidrug – resistant TB (MDR-TB) and Extensive Drug Resistant Tuberculosis (XDR-TB) are often fatal. Mortality rates of MDR-TB and XDR-TB are comparable with those for TB in the days before the development of antibiotics.

CHAPTER 4

DEADLY INTERACTION BETWEEN TUBERCULOSIS AND HIV/AIDS.

HIV/AIDS and TB are so closely connected that the term "co-epidemic" or "dual epidemic" is often used to describe their relationship. The intersecting epidemic is often denoted as TB/HIV or HIV/TB. HIV affects the immune system and increases the likelihood of people acquiring new TB infection. It also promotes both the progression of latent TB infection to active disease and relapse of the disease in previously treated patients. TB is one of the leading causes of death in HIV-infected people.

An estimate one-third of the 40 million people living with HIV/AIDS worldwide are co-infected with TB. Furthermore, without proper treatment, approximately 90% of those living with HIV die within months of contracting TB. The majority of people who are co-infected with both diseases live in sub- Saharan Africa.

Each disease speeds up the progress of the other, and TB considerably shortens the survival of people with HIV/AIDS. TB kills up to half up to half of all AIDS patients worldwide. People who are HIV-positive and infected with TB are up to 50 times more likely to develop active TB in a given year than people who are HIV-negative.

HIV infection is the most potent risk factor for converting latent TB into active TB, while TB bacteria accelerate the progress of AIDS infection in the patient. Many people infected with HIV in developing countries develop TB as the first manifestation of AIDS. The two diseases represent a deadly combination , since they are more destructive together than either disease alone.

TB is harder to diagnose in HIV-positive people. TB progresses faster in HIV-infected people. TB in HIV –positive people is almost certain to be fatal if undiagnosed or left untreated. TB occurs earlier in the course of HIV infection than many other opportunistic infections.

According to the World Health Organization (WHO) , TB infection is currently spreading at the rate of one person per second . It kills more young people and adults than any other infectious disease and is the world's biggest killer of women. WHO declared TB to be " a global health emergency" . Every year 8-10 million people catch the disease and 2 million die from it. About a third of the world's population, or around 2 billion people , carry the TB bacteria but most never develop the active disease.

Around 10% of people infected with TB actually develop the disease in their lifetimes, but this proportion is changing as HIV severely weakens the human immune system and makes people much more vulnerable.

Worldwide, women bear a disproportionate burden of poverty, ill-health, malnutrition and disease. TB causes more deaths among women than all causes of maternal mortality combined, and more than 900 million women are infected with TB worldwide. This year, 1 million women will die and 2.5 million mainly between the ages of 15 and 44 , will become sick from the disease.

Once infected with TB, women of reproductive age are more susceptible to developing TB disease than men of the same age. Women in this age group are also at greater risk of becoming infected with HIV. As a result , in certain regions, young women aged 15 -24 with TB outnumber young men of the same age with the disease.

While poverty is the underlying cause of much infection in rural areas, poverty is also aggravated by the impact of TB. A study by the World Bank, WHO and Harvard University reported TB as a leading cause of "healthy years lost" among women of reproductive age.

CHAPTER 5

POLICY FRAMEWORK FOR EFFECTIVE ACTION ON TUBERCULOSIS AND HIV/AIDS.

The deadly interaction of TB and HIV affects millions and threatens global public health. HIV has increased TB rates by as much as 500% in some countries of sub-Saharan African and urgent action is needed now to stop the co-epidemic. TB causes up to 50% of AIDS death in Africa. Two-thirds of people living with HIV in Africa lack access to effective TB diagnosis , prevention and treatment. Joint TB/HIV interventions can contribute to better TB control. TB/HIV collaboration can help in reaching the unreached and to get millions of people living with HIV on antiretroviral treatment.

TB control can contribute to better HIV/AIDS control both by reducing the TB burden in people with HIV and by providing an entry point to HIV prevention and care for people with TB. New resources are available, but are not accessible where they are most needed. Unprecedented global resources are being made available for AIDS and TB control, yet two-thirds of people living with HIV in sub-Saharan Africa lack access to DOTS treatment (the internationally recommended strategy for TB control). A combined approach could leverage additional resources in areas of greatest need.

Effective joint action is essential , and there is guidance from WHO on how best to facilitate this , including the Interim Policy on Collaborative TB/HIV Activities, the Strategic Framework to Decrease the Burden of TB/HIV , and Guidelines for Implementing TB and HIV programme activities. Advocacy and communications can make joint action more effective at global , regional and national levels by winning the support of key constituencies such as legislators , policy-makers and service providers in order to influence policies and spending and bring about social change.

Effective TB/HIV control requires committed political leadership, an uninterrupted supply of effective drugs, knowledgeable health workers and mobilized communities. Governments need to assign a high priority to TB control and HIV prevention and care, including increased collaboration between HIV and TB programmes. In many countries, TB control is a low political priority, and advocacy attempts need first to change the behavior of politicians, rather than risk groups or patients.

TB/HIV collaboration promotes a holistic approach to care that will reduce suffering among those affected by the dual epidemics. A combined approach can also reduce stigma, improve general health services, and strengthen civil society. In order to control TB, governments need to set up effective TB treatment programmes. The drugs and knowledge to control TB exists but the world's governments still need to wake up to the seriousness of the TB crisis and take action.

HIV infection is the most potent risk factor for converting latent TB into active transmissible TB- accelerating the spread of the disease- while TB bacteria help accelerate the progress of AIDS in HIV-positive people.

Joint TB/HIV interventions can contribute to better TB control, TB control can contribute towards better HIV control, and combining TB/HIV control can also lead to an improvement in general health services.

As HIV patients are more likely than others to develop active TB , new faster-acting ways to combat T5B are needed. Antiretroviral drugs can reduce TB by up to 80% in people with HIV. Today, TB is the leading cause of death in people who are HIV-positive . The two diseases represent a deadly combination – more destructive together than either is alone.

An effective and inexpensive cure for TB already exists , so the emphasis now must be on setting up more treatment programmes in more parts of the world. WHO is committed to getting people living with HIV on antiretroviral drugs. Significant misunderstanding of TB preventive therapy persists and substantial training and support are therefore essential before widespread implementation can be realistic.

The involvement of affected communities is needed at every stage of programmes to combat TB/HIV. Since the combination of diseases is deadly, joint approaches will be more effective than separate approaches . TB and HIV are often seen as medical problems , but this view limits the effectiveness of programmes. The more people who deliver the same message, the more difficult it will be for policy –makers to ignore.

To implement or enhance community-based care for TB/HIV patients and TB preventive therapy in HIV patients, training must target national and district policy-makers, local leaders , health care workers , community workers , volunteers, patients and family members.

People with TB and/or HIV often have a range of conditions and should not need to attend health services separately for each of them. Access to diagnosis and treatment of TB/HIV is a human rights issue and people should have a right to treatment if they have TB or HIV. The World Health Organization Interim Policy on Collaborative TB/HIV Activities provides national governments and managers of both TB and HIV programmes with guidance on addressing the dual epidemic of TB and HIV. The policy provides a road map for expanding collaboration between national TB and HIV/AIDS programmes to curb the pandemic of co-infection.

The policy advice is aimed at decision- makers in the field of health and TB and HIV programme managers working both in the public health field and in other sectors, as well as donor agencies, development agencies and nongovernmental organizations supporting TB and HIV programmes.

The Global TB/HIV Working Group, which coordinates the global response to the intersecting TB and HIV epidemics, has formulated the policy . Its membership includes programme managers , development agencies , nongovernmental organizations, academic institutions, activists and patient-supporting groups working with WHO and UNAIDS on both TB and HIV programmes. The writing committee included technical experts from TB and HIV, policy-makers involved in health management , persons living with HIV and their advocates, international and national TB and HIV programme managers, and donor agencies.

The policy offers direction on which collaborative TB/HIV activities to implement and the circumstances in which they should be implemented. These activities are complementary to , and in synergy , with the established core activities of TB and HIV prevention and control programmes.

Implementing the DOTS strategy is the core activity for TB control. Similarly, infection and disease prevention , health promotion activities and the provision of treatment and care for the basis for HIV control. The policy does not call for the institution of a new specialist or independent disease control programme. Rather, it promotes enhanced collaboration between TB and HIV programmes in the provision of a continuum of high-quality care at service- delivery level for people with or at risk of TB and people living with HIV.

The objectives of collaborative TB/HIV activities are to establish the mechanisms for collaboration between TB and HIV/AIDS programmes, to reduce the burden of

TB in people living with HIV/AIDS ; and to reduce the burden of HIV in TB patients.

CHAPTER 6

ESTABLISHMENT OF MECHANISMS FOR COLLABORATION IN TB/HIV/AIDS ACTIVITIES.

In the establishment of mechanisms for TB and HIV/AIDS collaboration, there is need to set up a coordinating body for TB/HIV activities effective at all levels. HIV/AIDS and TB programmes should create joint national TB and HIV coordinating bodies at regional , district and local levels (sensitive to country-specific factors), with equal or reasonable representation of two programmes, including TB and HIV patient support groups.

Surveillance of HIV prevalence among tuberculosis patients should be conducted. There should be HIV prevalence among TB patients in all countries, irrespective of national adult HIV prevalence rates. Countries with unknown HIV prevalence rates among TB patients should conduct a seroprevalence (periodic or sentinel) survey to assess the situation. In countries with a generalized epidemic state , [2] HIV testing and counselling for all TB patients should form the basis for the surveillance. If this is not yet in place, periodic surveys or sentinel surveys are suitable alternatives.

In countries with a concentrated epidemic state, where HIV prevalence is consistently >5% in at least one defined subpopulation and is <1% in pregnant women in urban areas, where groups at high risk for HIV are localized in certain administrative areas, HIV testing and counselling for all TB patients in those administrative areas should form the basis for the surveillance. If this is not yet in place, periodic surveys or sentinel surveys are suitable alternatives.

In countries with a low-level epidemic state , where HIV prevalence has not consistently exceeded 5% in any sub population, periodic surveys or sentinel surveys are recommended.

There should be joint planning for TB/HIV activities in the areas of resource mobilization , capacity building, including training, advocacy, programme communication , social mobilization, enhancement of community involvement and operational research. There should also be collaborative monitoring and evaluation of TB/HIV activities.

Countries should ensure mobilization of sufficient and qualified human resources to implement collaborative TB/HIV activities in accordance with country- specific situations. The TB/HIV coordinating bodies should be responsible for the governance and mobilization of resources to implement collaborative TB/HIV activities , thereby avoiding competition for the same resources.

TB and HIV/AIDS programmes should draw up a joint training plan to provide pre-service and in-service training and continuing medical education on collaborative TB/HIV activities for all categories of health care workers.

Tuberculosis and HIV/AIDS programmes should ensure that the capacity of the health care delivery (e.g. laboratory, drug and referral capacity) is adequate for effective implementation of collaborative TB/HIV activities.

Well designed TB/HIV advocacy activities , jointly planned to ensure coherence of their messages and targeted at key stakeholders and decision- makers , should be carried out at global, national, regional and local levels.

HIV/AIDS and TB programmes should develop joint TB/HIV programme communication and social mobilization strategies, which address the needs of individual clients and patients and of communities affected by HIV/AIDS and TB.

The joint communication strategies should ensure the mainstreaming of HIV communication components in TB communication and of TB communication components in HIV communication.

All stakeholders, including HIV/AIDS and TB programmes , should ensure the involvement of TB and HIV patient support groups and their communities in the planning , implementation and advocacy of collaborative TB/HIV activities.

All stake holders of collaborative TB/HIV activities should support and encourage TB/HIV operational research on country- specific issues to develop the evidence base for efficient and effective implementation of collaborative TB/HIV activities.

HIV/AIDS and TB programmes should agree on a core set of indicators and data collection tools , and collect data for monitoring and evaluation of collaborative TB/HIV activities.

The WHO guideline for monitoring and evaluation on collaborative TB/HIV activities should be used as a basis to standardize country specific monitoring and evaluation activities.

CHAPTER 7

REDUCTION OF TB BURDEN ON PEOPLE LIVING WITH HIV/AIDS (PLWHA)

Intensified TB case –finding should be established in all HIV testing and counselling settings using, at a minimum, a simple set of questions to identify suspected TB cases as soon as possible. The questions should be asked by trained counsellors.

A referral system should be established between HIV counselling and testing and TB diagnostic and treatment centers. TB case-finding in people living with HIV/AIDS (PLWHA) in clinics and hospitals , household contacts, populations at high risk of HIV, and congregate settings should be intensified by increasing the awareness and knowledge of interactions between TB and HIV in health care workers and the populations they serve, identifying suspected TB cases and referring them for diagnosis , on a regular basis.

HIV/AIDS programmes should provide Isoniazid Preventive therapy (IPT) as part of the package of care PLWHA when active TB has been safely included. Information about IPT should be made available to all PLWHA.

Each health care and congregate setting should have , and implement , a TB infection control plan, supported by all stake holders, that includes administrative , environmental and personal protection measures to reduce transmission of TB.

CHAPTER 8

REDUCTION OF HIV BURDEN IN TB PATIENTS.

HIV testing and counselling should be offered to all TB patients in settings where the HIV prevalence among TB patients exceeds 5%. TB control programmes should mainstream provision of HIV testing and counselling in their operations or establish a referral linkage with the HIV/AIDS programmes for this purpose. All clients attending TB clinics should be screened for sexually transmitted infections (STIs) using a simple questionnaire. Those with symptoms of STIs should be treated or referred to STI treatment providers.

TB control programmes should implement procedures for reduction of occupational and nosocomial exposure to HIV in their services. TB control programmes should provide harm reduction measures for TB patients when injecting drug use is a problem or establish a referral linkage with HIV/AIDS programmes for this purpose.

TB control programmes should ensure that vertical transmission is prevented by referring pregnant HIV-infected clients to providers of services for prevention of mother-to- child transmission. TB/HIV/AIDS programmes should establish a system for providing Co-trimoxazole preventive therapy (CPT) to eligible PLWHA who have active TB.

All PLWHA who are dignosed with TB should also be provided with HIV/AIDS care and support service. TB control programmes should establish a referral linkage with HIV/AIDS programmes to provide a continuum of care and support for PLWHA who are receiving or have completed their TB treatment.

Antiretroviral therapy (ART) should be offered to all HIV-positive TB patients depending on the eligibility criteria for ART in TB patients in each country and the drug interactions (with rifampicin) . TB and HIV/AIDS programmes should create the mechanism to provide ART to eligible HIV-positive TB patients.

From the foregoing, It is therefore recommended that before starting collaborative TB/HIV activities, that countries with national adult HIV prevalence rate ≥ 1% should implement all collaborative TB/HIV activities described above, countries with national adult HIV prevalence rate ≥ 1% should implement all collaborative TB/HIV activities in those administrative areas with adult HIV prevalence rate ≤ 1% and should implement activities in other parts of the country while countries with national adult HIV prevalence rate below 1% should undertake surveillance of HIV prevalence among TB patients and implement the activities aimed at reducing the burden of TB in PLWHA (Intensified TB case-finding , isoniazid preventive therapy, and TB infection control in health care and congregate settings).

CHAPTER 9

TUBERCULOSIS AND HIV/AIDS INTERNATIONAL DAYS.

If the dual epidemic of TB and HIV is to be tackled effectively , it is important for established AIDS programmes to integrate TB messages into their campaigns and to mainstream joint TB/HIV messages into ongoing AIDS advocacy. Examples of this can include HIV organizations incorporating messages and information about TB in campaign materials, health promotion literature, websites and policy advice.

Equally, TB programmes need to make further connections with HIV and AIDS messages and organizations and seek to include TB/HIV policy –related activity in their own work.

International events are a valuable opportunity to raise awareness about the state of TB/HIV in the world today as well as its prevalence and impact on national and regional levels.

The World TB Day held on 24th of   March each year , is an occasion for people around the world to raise awareness about the International health threat presented by TB. It is a day to recognize the collaborative efforts of all countries involved in fighting TB. TB can be cured , and, with diligent efforts and sufficient resources, eventually eliminated.

On 24th March 1882, Dr. Robert Koch announced the discovery of the TB bacillus. In 1982, a century later, the first World TB Day was sponsored by the World Health Organization and the International Union Against Tuberculosis and Lung Disease.

The theme and slogan for the 2004 campaign , aimed principally at the general public and the media is "Every Breath Counts, Stop TB Now!' There is an inextricable link between the act of breathing and life itself. Breath and Breathing are also closely associated with TB. Although World TB Day is a worldwide event, different countries and regions choose locally related activities and messages. This global call to action is also a way to mobilize political and social commitment.

The International AIDS Candlelight Memorial Campaign involves all sectors of the local community in the fight against HIV/AIDS; each year. Memorials take place in more than 1500 communities scattered over more than 85 countries.

On Sunday, 16th May 2004, communities around the world came together in solidarity to light candles and remember those who have been touched by HIV/AIDS. Local events are coordinated by organizations , individuals, governments and faith-based communities.

The first international AIDS Candlelight Memorial was held in 1983, when the cause of AIDS was unknown and no more than a few thousand AIDS deaths had been recorded. The organizers wished to honour the memory of those who had died and demonstrate support for those living with AIDS.

The World AIDS Day on every 1st of December each year marks progress made in the battle against the epidemic and bring into focus remaining challenges including the undeniable need for greater joint TB/HIV programming.

World AIDS Day helps to raise awareness, political commitment and resources for the global effort. It is also a major focus for prevention activities , education and fighting prejudice. World AIDS Day reminds the world that HIV has not gone away, and that much remains to be done.

CHAPTER 10

TUBERCULOSIS, HIV/AIDS AND WOMEN.

HIV prevalence in women is highest in developing countries . This means that many women are also at serious risk from TB if services are not made more easily accessible to them. TB is recognized as the single biggest infectious killer of women in the world, and accounts for more cases of maternal mortality than all other causes put together . More than one million women die needlessly from TB every year.

Data from WHO confirm that TB is the leading cause of death among women of reproductive age. More than 900 million women are infected with TB worldwide and TB accounts for 9% of deaths worldwide among women aged between 15 and 44.

In 2002, about half of all people living with HIV or AIDS worldwide were female and in sub-Saharan Africa, home to 70% of all HIV-infected people. More than half of all infected adults are women .

Women between the ages of 15 and 24 in settings of high HIV prevalence are increasingly more likely than men in the same age group to fall sick with TB. Women in this age group are also at greater risk from HIV infection. Among women sick with TB , at least a third die because they are undiagnosed or receive poor treatment.

Since TB affects women mainly in their economically and reproductively active years, the disease also has a heavy impact on their children and families.

Children are at risk of contracting TB, because of close contact with their mothers.

Although global TB detection rates appear to be higher in men than in women, this may not reflect the real situation. Lower prevalence rates of TB among women , particularly in the 15-35 age group, may be the result of under notification of infected women.

A combination of various cultural, social and economic factors, especially in low –income countries, means that women often face difficulties in accessing health care. Therefore, by the time they attend clinics , TB and/or HIV can already be at a very advanced stage.

The stigmatization of women who are HIV-positive or who have TB discourages them from seeking treatment. Women have to overcome several barriers before they can easily access health care services. They are often unable to leave their home and work or they may need permission from their families to go to a clinic or to pay for treatment.

Women have long been key elements in the care and management of both TB and AIDS , not only in their own families but also in the wider community. Key features of women- only services are easy access, social support, and appropriate advice and treatment. Facilities should be women –centered, women- managed-and offer affordable , accessible services.

On a global level, programmes for TB control should respond to the special needs of women in order to promote health and to reduce possibly unequal access to health care.

The Global Coalition on Women and AIDS, launched in February 2004, is an informal grouping of partners and organizations working to mitigate the impact of AIDS on women and girls worldwide. It is a growing global , inclusive movement seeking to support, energize and drive AIDS-related programmes and projects to improve the daily lives of women and girls. The coalition aims to build global and national advocacy to highlight the effects of HIV and AIDS on women and girls and to stimulate concrete, effective action. Efforts are focused on preventing new HIV infections, promoting equal access to treatment, addressing legal inequities and mitigating the impact of AIDS on women and girls. The Coalition brings technical expertise and spokespeople to the World AIDS Day activities. [2]

Women, Girls and AIDS was the theme for World AIDS Day 2004. World AIDS Day marks progress made in the battle against the epidemic and brings into focus remaining challenges, one of which is undeniably the need for greater joint TB/HIV programming.

World AIDS Day helps to raise awareness , political commitment and resources for the global effort. It is also a major focus for prevention activities, education and fighting prejudice. World AIDS Day reminds the world that HIV has not gone away , and that many things remain to be done. [3]

CHAPTER 11

TUBERCULOSIS AND HIV/AIDS WORKSHOPS.

Key issues raised at TB/HIV Workshops like the Stop TB Partner's Forum has objectives to identify advocacy and mobilization issues around TB/HIV. Some of these workshops are coordinated by the Global Network of People Living with HIV/AIDS (GNP+) and the communications focal point for infected and affected communities on the board of the Global Fund to Fight AIDS, Tuberculosis and Malaria.

Participants are introduced and reminded to seek to " Begin to identify key issues and opportunities in advocacy and community mobilization around TB/HIV". Participants focus on advocacy and community mobilization and resist the temptation to concentrate too much on the challenges around the programmatic integration of TB/HIV action.

Participants introduce themselves and are invited to describe their current responsibilities and influence and also, if they wished , a little of their own personal TB/HIV history. The range of background and involvement are diverse, with several participants choosing to share their own experiences of either TB or HIV.

CHAPTER 12

OVERVIEW OF TB AND HIV PERSPECTIVE BY WORLD HEALTH ORGANIZATION.

Presentations on the TB/HIV Working Group of Stop TB focus on three main aspects of TB/HIV ; the rationale for action, the necessary response, and the advocacy needed. Key points from these presentations include the rationale for the DOTS strategy for fighting TB, it's necessity and sufficiency in countries with high HIV prevalence. An estimated 37% of all TB deaths in sub- Saharan Africa are related to HIV, many countries ignore the recommendation on isoniazid preventive therapy, TB patients need to know their HIV status, mortality from TB ranges from 11% to 50% for HIV+ people, TB is often not identified in HIV + people and is therefore very dangerous , as the infection moves quickly.

Response by WHO and the Stop TB partnership established a TB/HIV Working Group to explore and help define key policy arguments , and develop a strategic framework and guidelines for developing countries . [4] A number of " ProTEST" projects have been running and have helped to establish a strong evidence base for further action. Malawi is a leader on TB/HIV and aim to deliver ARVs with the same philosophy, networks and infrastructure as anti- TB drugs.

There is a need for advocacy work at three levels; country level, to support joint TB/HIV programme activities through advocacy of civil society and groups of

PLWHA, policy level , aimed at national level, to create the environment for effective joint TB/HIV activities and to access resources; global level, which at the moment is the weakest area and needs further development.

CHAPTER 13

KEY ISSUES IN TB/HIV ADVOCACY AND COMMUNITY MOBILIZATION.

In these TB/HIV workshops, key issues in TB/HIV advocacy and community mobilization are identified. It will be felt that there is a significant clash of cultures between TB and HIV that needed to be acknowledged in taking forward any successful advocacy and community mobilization.

There is need to know what it means for TB to engage in the "messiness " of activism and to know whether TB public health experts really thought this through.

There is need to know how the very different philosophies as reflected in their key images – " It is outrageous to die of AIDS", as opposed to " Let's control TB" will be reconciled.

TB should seek to build on HIV experiences for effective advocacy as HIV advocacy is perceived as a success.

It is essential to identify an existing political platform from which to launch the issue of TB/HIV . Successful HIV activism was built in part on the foundations of the gay movement in the West and on the anti-apartheid movement in South Africa. There is need to urgently identify platforms to be used for TB/HIV.

People with HIV significantly set the agenda for the AIDS fight but this is not the case for TB . It was strongly felt by many that HIV is owned by activists and TB owned by public health professionals.

TB is an "opportunistic " infection- the advocacy too should be "opportunistic.

More "passion" - rather than "word-perfect" messaging,- is desperately needed in TB advocacy. We need a new thinking to understand the reality for TB. Identity was felt to be an issue : HIV-for life, TB= cured. There is a need for louder voices for patients through the media.

On the resource implications on timing, there is only a short window of opportunity , in light of combining any work with the increased roll out of ARVs, to support this work and get it mobilized.

There is an urgent need to expand and mobilize a " real TB community" . "TB community= a professional community " . It was felt by some that TB is " owned" by well- meaning public health officials. The need to open up the TB community to create greater dynamism was cited. TB support/activism must make access to information easier on TB /HIV to facilitate activism. Building capacity around advocacy action and TB is an urgent need. "Ownership " of response should be transferred from Stop TB and WHO , for example by encouraging more TB patient groups and training them in advocacy skills such as use of the media to highlight particular issues. We should encourage a human right approach to treatment. We should determine if it is an activist role or a public health role and also find out the best role for public health.

It was felt that stigma is manifested differently in HIV and TB and needs to be addressed through advocacy efforts and programme communications. In many countries, the stigma that attached to death means that " died of TB " is better than "died of AIDS".

More education on TB/HIV is needed for those working on TB and HIV . The stigma of TB can also extend to TB and poverty. Much stigma persists within the medical community itself and needs to be tackled- for example, some working in the field have the impression that DOTS programmes are disinclined to take on HIV+ individuals whose potential lack of response to DOTS might adversely affect statistical outcomes.

More programme links are needed and advocacy can help facilitate this. In the United Kingdom , as in many other countries , TB is the most common AIDS-defining illness but health care services often miss the links. There was also a suggestion of ignorance among health professionals of possible drug interactions. Putting TB systems onto existing HIV systems for both care and advocacy will be very useful.

The priority is to scale up TB case detection and treatment. There is need to raise TB/HIV issues at country level among policy –makers to support programme implementation. It is very important to know how this should be done.

Many people , including those living with HIV, do not know that TB is curable. This is a basic message and one thing that is failing to get through. We should know how to get the message to policy- makers and funders that more drugs are needed for both diseases. It is not "either/or', there is no point providing ARVs only for patients to die of TB for lack of TB resources. Similarly , TB efforts alone are insufficient.

The need for the right messages about "treatment" of TB/HIV, particularly about ARV rollout and prophylaxis for TB is crucial. There was some criticism of the emerging "three ones" approach in HIV, which sets out to bring all efforts under one national strategy. It was felt that success needs many voices to push the agenda and cannot /should not always be coordinated. Those working in TB should recognize that much of the success in driving forward the response to AIDS has been facilitated through diverse messages from very different perspectives.

Synergies must be identified in the dual epidemics, just as with intravenous drug-users and HIV prevention and care. There is need to understand whose advocacy voice should be used . The messaging and packaging from Stop TB is good but there is also an urgent need for messaging that really challenges that status quo and shifts the response.

Advocacy does not need one message and should not always look to integrate action. An effective response needs many voices. WHO/Stop TB need to be prepared to let go and prepare for a tough time to achieve the ultimate goals.

CHAPTER 14

KEY OPPORTUNITIES FOR ADVOCACY AND COMMUNITY MOBILIZATION.

Participants in TB/HIV/AIDS workshops will be asked to identify opportunities for advocacy and community mobilization. There are some good existing resources that should be used. For example, a report on TB/HIV produced by the Results Educational Fund in the USA could be used as a basis for strong advocacy. There is need for an advocacy e-forum discussion to get the dialogue moving. The Bangkok Conference offers an excellent communication opportunity to mobilize the TB activist movement using plenary session, non-abstract sessions, synergies for TB/HIV and ARV delivery , skills building , etc.

The GFATM partners forum in Bangkok offers similar excellent opportunities. Passion must be heard from Stop TB and the TB community at the Bangkok conference –not messaging but passion.

Businesses should be targeted . A recent survey shows that 80% of business leaders feel that HIV is bad for business , but only 4% feel TB is bad for business. Business is just one area that needs to be targeted.

Positive Nation (the leading United Kingdom magazine for those living with HIV) agreed to continue to publish feature material on TB/HIV. The World Health Assembly was also identified as an opportunity to raise TB/HIV attempts to get a focus session but was not successful but TB/HIV would be raised in the context of other programmes such as 3 x 5 and ARV rollout. There will be further specifications in the future.

The media opportunities of World TB Day raise issues in the Stop TB Partners' Forums. UNAIDS/Stop TB are developing further media and communications support around the subject , and a TB/HIV section of the UNAIDS Bangkok report is planned with associated communications report.

TB/HIV advocacy workshops should be organized by UNAIDS and Stop TB to build on advocacy sessions. A joint UNAIDS/StopTB advocacy pack on TB/HIV should be finalized and distributed in advance before conferences to many AIDS organizations and TB organizations, and published on the UNAIDS and Stop TB websites.

TB/HIV advocacy should be raised as an issue for further considerations in all the organizations. Messaging and advocacy opportunities will be greatly increased through links that will get many more people on ARVs, if large numbers of co-infected people can be diagnosed and cured of TB before starting ARV therapy, or treated for both diseases at the same time . There should be increased advocacy activity to promote GFATM and the programmes it supports, particularly the representative on the GFATM board for infected and affected communities.

CHAPTER 15

TUBERCULOSIS AND HIV/AIDS INITIATIVES.

TB/HIV/AIDS initiatives target to get millions of people in developing and middle-income countries on antiretroviral treatment (ART). Globally, 40 million people are infected with HIV/AIDS. Every single day AIDS kills 8000 people and orphans thousands of children. Heavily affected countries face total social and economic collapse within just a few generations if decisive steps are not taken.

Treatment exists that can keep people alive and transform HIV/AIDS from a death sentence to a manageable chronic disease. Until now, however, treatment has been the most neglected area of HIV/AIDS programming.

The treatment gap facts are that 6 million people desperately need treatment, 3million die every year because they cannot get the necessary drugs, only 400,000 people have access to treatment worldwide and Africa is home to 70% of people with HIV – less than 2% of those in need have access to ART.

The failure to deliver life-prolonging drugs to millions of people in need was declared a global health emergency at UNGASS in September 2003. Less than 3 months later, on World AIDS Day 2003, WHO and UNGASS launched the "3 by 5" initiative – an ambitious target to get 3 million people living with AIDS on antiretroviral treatment by the end of 2005. This target is a vital step towards the ultimate goal of providing universal access to AIDS treatment to all those who need it.

CHAPTER 16

AIDS MEDICINES AND DIAGNOSTICS SERVICES (AMDS)

The WHO and UNAIDS "3 by 5" strategy focuses on providing developing countries with support, in the form of simplified norms and guidelines and other forms of direct technical assistance, for scaling up antiretroviral therapy. Because procurement and supply chain management of pharmaceuticals and diagnostics are significant problems for most poor countries, WHO has established the AIDS Medicines and Diagnostics Service (AMDS) to help such countries with all aspects of selecting , procuring and delivering both HIV medicines and diagnostic tools to the point of service delivery.

The WHO and UNAIDS strategy is based on five key pillars; global leadership , strong partnership and advocacy, urgent and sustained country support; simplified , standardized tools for delivering ART; effective , reliable supply of medicines and diagnostics; rapidly identifying and reapplying new knowledge and successes. [5]

The '3 by 5" movement operates within the Three Ones framework , a concept adopted by donors and aid agencies to enhance coordination on the ground and thereby avoid duplication . The framework calls for one HIV/AIDS action framework to provide the basis for coordinating the work of all partners; one national AIDS coordinating body, with a broad-based multi-sectoral mandate; and one country level monitoring and evaluation system.

The highlights of progress to date includes; development of simplified antiretroviral drug regimens, and testing and treatment guidelines that are consistent with the highest standards of care, harmonized global monitoring and evaluation framework for ART programmes , with patient –tracking tools in development, launch of streamlined guidelines for training health workers in a wide range of skills , from HIV counselling and testing and recruitment of patients to treatment delivery, treatment modules in integrated management of adult and adolescent illness, clinical management of patients and monitoring of drug resistance.

 In partnership with UNICEF and the World Bank , there was establishment of the AIDS medicines and diagnostics service to ensure that developing countries have access to quality antiretroviral drugs and diagnostic tools at the best prices, pre-qualification of both brand- name and generic drugs, including single- drug, two-drug and three- drug fixed –dose combinations (FDCs) according to stringent standards of quality , safety and efficacy. These FDCs are saving lives now in many countries.

Close to 50 countries have appealed to WHO for urgent support in establishing or scaling up ART programmes; to date , WHO staff have visited 28 of these countries (including the Russian Federation and Ukraine). A total of 40 country coordinators have been recruited , followed by additional staff , to cover 40 countries in support of scale-up.

CHAPTER 17

ANTIRETROVIRAL THERAPY (ART) TREATMENT AND PREVENTION PLAN.

ART prolongs lives, making HIV/AIDS a chronic disease- not a death sentence. Affluent countries have seen a 70% decline in HIV/AIDS deaths. ART helps to calm fears and change attitudes towards HIV. As part of the prevention plan, ART can significantly reduce HIV transmission. ART can reduce overall health care costs and restore quality of life.

WHO and UNAIDS are working to make ART accessible to all.

To ensure a comprehensive response to HIV/AIDS, treatment and prevention programmes must enhance and accelerate each other. When people have hope that they can be treated and lead productive lives, their desire to know their status and to protect themselves and their partners is much greater. Evidence and experience show that rapidly increasing the availability of ART increases community awareness about HIV/AIDS, promotes uptake of HIV testing and can lead to more openness about AIDS. Individuals receiving effective treatment are also likely to be less infectious and less able to spread the virus.

CHAPTER 18

ESTABLISHMENT OF GLOBAL PARTNERSHIPS TO STOP TUBERCULOSIS .

The Global Partnership to Stop TB is a global movement to accelerate social and political action to stop the spread of tuberculosis around the world. The Stop TB mission is to increase access , security and support in order to ensure that every TB patient has access to TB treatment and cure, and protect vulnerable populations from TB, reduce the social and economic toll that TB exacts from families, communities , and nations . The Partnership's approach is a coordinated , multinational , multisectoral global effort to control TB.

Partnership and collaboration between other United Nations agencies and multilateral agencies, NGOs, foundations, community organizations , faith-based organizations , the HIV activist community , the private sector, trade unions and representatives of the community of people living with HIV/AIDS are absolutely essential if " 3 and 5" to be accomplished. '3 by 5" has already given rise to unprecedented interest in collaborating with WHO and UNAIDS and/or contributing to the target in some way, and WHO continues to build these relationships and partnerships.

Successive examples in Brazil and pilot projects in other countries have shown that increasing access to treatment is both possible and effective. Brazil has the most advanced national HIV/AIDS treatment programme in the developing world, averting almost 100,000 deaths - a 50% drop in AIDS mortality per year.

Results in Brazil clearly demonstrate how scaling up can also help strengthen health systems and dramatically reduce public health costs. The programme has brought about a significant decline in the number of hospital admissions and cost savings in reduced admissions and opportunistic infections are estimated at more than US$ 1 Billion.

The programme has also been effective in reducing the rates of TB and other opportunistic infections. [6]

CHAPTER 19

THE DIRECTLY OBSERVED TREATMENT SHORTCOURSE (DOTS) STRATEGY.

The internationally recommended strategy to control TB , known as DOTS , has five components which are to establish political commitment to sustained TB control, access to quality –assured TB sputum microscopy, standardized short-course chemotherapy, including direct observation of treatment, an uninterrupted supply of drugs and a standardized recording and reporting system , enabling assessment of outcome of patients.

CHAPTER 21

CONCLUSION

HIV/AIDS is dramatically fuelling the TB epidemic in sub-Saharan Africa, where up to 70% of TB patients are co-infected with HIV in some countries. For many years efforts to tackle TB and HIV have been largely separate, despite the overlapping epidemiology. Improved collaboration between TB and HIV/AIDS programmes will lead to more effective control of TB among HIV- infected people and to significant public health gains.

The World Health Organization's interim policy on collaborative TB/HIV activities [1], gives guidance on what should be done to address the dual TB and HIV epidemic. This includes the identification of collaborative TB/HIV activities and the establishment of TB/HIV coordinating bodies to promote and coordinate the response of the two programmes at all levels.

HIV- positive people can easily be screened for TB; if they are infected they can be given prophylactic treatment to prevent development of the disease or curative drugs if they already have the disease.

TB patients can be offered an HIV test ; indeed , research shows that TB patients are more likely to accept HIV testing than the general population. This means TB programmes can make a major contribution to identifying eligible candidates for ARV treatment.

This is a crucial moment in the history of HIV/AIDS and an unprecedented opportunity to alter its course. The international community has the chance to change the history of health for generations to come and to open the door to better health for all.

BIBLIOGRAPHY.

1. World Health Organization Report . *Interim policy on collaborative TB/HIV activities*. World Health Organization, Geneva, 2004. (WHO/HTM/TB/2004.330 and WHO/HTM/HIV/2004.1)
2. The Global Coalition on Women and Aids. http://www.womenandaids.org.
3. World AIDS Day. http://www.unaids.org.
4. Strategic framework to decrease the burden of TB/HIV . Geneva World Health Organization, 2002 (WHO/CDS/TB/2002.296). Available from WHO at cdsdoc@who.int.
5. National AIDS Drug Policy. Brasilia, Ministry of Health of BRAZIL, 2002.
6. The World Health Report 2004; Changing History; Http;//www.who.int/whr/en.

www.ingramcontent.com/pod-product-compliance
Lightning Source LLC
Chambersburg PA
CBHW021851170526
45157CB00006B/2395